Project AIR F

HOW SH
THE U.S.
AIR FORCE
DEPOT
MAINTENANCE
ACTIVITY GROUP
BE FUNDED?

INSIGHTS FROM
EXPENDITURE AND
FLYING HOUR DATA

EDWARD G. KEATING
FRANK CAMM

Prepared for the
UNITED STATES AIR FORCE

RAND

The research reported here was sponsored by the United States Air Force under Contract F49642-01-C-0003. Further information may be obtained from the Strategic Planning Division, Directorate of Plans, Hq USAF.

Library of Congress Cataloging-in-Publication Data

Keating, Edward G. (Edward Geoffrey), 1965–
    How should the U.S. Air Force Depot Maintenance Activity Group be funded? : insights from expenditures and flying hour data / Edward G. Keating, Frank Camm.
        p. cm.
    "MR-1487."
    Includes bibliographical references.
    ISBN 0-8330-3143-0
    1. Depot Maintenance Activity Group (U.S.) 2.United States. Air Force—Equipment—Maintenance and repair. 3. United States. Air Force—Operational readiness. 4. United States. Air Force—Appropriations and expenditures. I. Camm, Frank A., 1949– II. Title.

UG1103 .K43 2002
358.4'175'0973—dc21

                                        2002017938

Published 2002 by RAND
1700 Main Street, P.O. Box 2138, Santa Monica, CA 90407-2138
1200 South Hayes Street, Arlington, VA 22202-5050
201 North Craig Street, Suite 202, Pittsburgh, PA 15213-1516
RAND URL: http://www.rand.org/
To order RAND documents or to obtain additional information, contact Distribution Services: Telephone: (310) 451-7002;
Fax: (310) 451-6915; Email: order@rand.org

Despite much effort and high leadership visibility over the last several years, the U.S. Air Force continues to experience unacceptably low mission capability (MC) rates for critical weapon systems and unexpectedly high support costs for these weapon systems at the end of each fiscal year. Multiple reviews over the last several years of the logistics process that the U.S. Air Force uses to support its weapon systems all agree that the process does not work well in its current environment.[1] It needs to change to meet the needs of a force and procedures that are different from those in place when the process was devised. Different reviews find different sources of problems. Taken together, the reviews point to two kinds of problems:

- Given the resources it is willing to commit to logistics activities and how its logistics process actually performs, the Air Force tries to do more with its operational weapon systems than its logistics support budgets allow.

---

[1]See, for example, Air Force Materiel Command Reparable Spares Management Board (Frank Camm, chair), *Final Report,* Wright-Patterson Air Force Base, Ohio, March 1998. The following General Accounting Office reports offer a variety of corroborating empirical evidence on the recent state of the Air Force depot management system: *Air Force Supply Management: Analysis of Activity Group's Financial Reports, Prices, and Cash Management,* GAO/AIMD/NSIAD-98-118, June 1998; *Air Force Supply: Management Actions Create Spare Parts Shortages and Operational Problems,* GAO/NSIAD/AIMD-99-77, April 1999; *Air Force Depot Maintenance: Analysis of Its Financial Operations,* GAO/AIMD/NSIAD-00-38, December 1999; and *Air Force Depot Maintenance: Budgeting Difficulties and Operational Inefficiencies,* GAO/AIMD/NSIAD-00-185, August 2000.

- Given the resources it is willing to commit to logistics activities, the Air Force could improve how its logistics process performs.

The Air Force Directorate of Supply (AF/ILS) asked RAND to examine the first kind of problem. When AF/ILS began the Spares Campaign in March 2001, it asked RAND to study particular issues. The RAND team focused on improving procedures relevant to programming and budgeting decisions for the depot-level logistics process for the following reasons:

- Programming and budgeting decisions are related to problems associated with the level of funding available for logistics activities designed to meet set weapon system availability targets.

- No matter how the Air Force tries to improve its logistics process, it will encounter adverse weapon system availability and cost outcomes unless its logistics programs and budgets reflect how the logistics process will actually work during the program and budget period. Getting these decisions right is a precondition to resolving other problems in the supply chain.

- Prices in working capital funds, which have been implicated in significant disconnects between users and providers of depot-level logistics services, are an integral part of the Air Force budgeting process. Any real change in the logistics budgeting process must address these disconnects.

- Success in using empirically based evaluation to close the loop on programming and budgeting decisions could point the way to better organizational relationships and better analytic approaches, with implications for helping to close the loop on other decisions as well.

This report addresses a key element of the analysis undertaken in this project. The Air Force Planning, Programming, and Budgeting System (PPBS) process today ultimately asks operating commands to develop budgets for the depot-level support they will need. Such an approach naturally raises the following question: How should the level of activity in an operating command in a particular future year addressed in the PPBS process be related to the actual costs that Air Force Materiel Command (AFMC) will have to cover in that year? The Air Force has evolved a system, commonly known as the

"AFCAIG process," to make this link. The Air Force Cost Analysis Improvement Group (AFCAIG) coordinates the effort. But each year over the last decade, the money that the operating commands have provided to cover AFMC costs has fallen short of costs. This outcome is an integral part of the problem of "unexpectedly high support costs" mentioned in the first paragraph.

To provide an analytic context for understanding this problem, the study team sought a basic understanding of the link between operating command activities and actual AFMC costs. As explained in the text, the analysis focused on one key link to allow some precision: how costs in depot-level maintenance shops funded by the Depot Maintenance Activity Group (DMAG) vary in response to changes in flying hours in the operating commands.

This report should interest those responsible for planning, analyzing, or improving budgeting and pricing systems that coordinate the activities of separate government organizations. It should be of particular interest to those working on policy for defense working capital funds that buffer the transfers between one defense organization incurring costs to provide a service and another consuming and funding the service. The research took place in Project AIR FORCE's Resource Management Program.

PROJECT AIR FORCE

Project AIR FORCE, a division of RAND, is the Air Force federally funded research and development center (FFRDC) for studies and analyses. It provides the Air Force with independent analyses of policy alternatives affecting the development, employment, combat readiness, and support of current and future aerospace forces. Research is performed in four programs: Aerospace Force Development; Manpower, Personnel, and Training; Resource Management; and Strategy and Doctrine.

CONTENTS

FIGURES

TABLES

The basic inquiry in this report is how Air Force Materiel Command (AFMC) depot-level expenditures relate to operating command activity levels, i.e., flying hours. We hypothesize that a large portion of depot-level costs are unrelated to operating command activities.

APPROACH

We examine the recorded expenditures of AFMC's Depot Maintenance Activity Group (DMAG) and relate Mission Design (MD)-specific DMAG repair expenditures to various lags of fleet flying hours.

Our analysis uses H036A data, which track DMAG expenditures by Job Order Number (JON). The data describe where work was performed and for which MD and customer; the data break expenditures into various types, e.g., direct civilian labor, materiel, overhead. Direct civilian labor is a minority of organic DMAG expenditures; considerably more is spent on materiel and operating overhead.

The basic model we estimated was a linear regression with DMAG expenditures as the dependent variable and various lags of fleet flying hours the independent variables. If increasing flying hours increases DMAG expenditures, we expect positive regression coefficient estimates.

LINKING DMAG EXPENDITURES TO OPERATING COMMAND FLYING HOURS

The DMAG directly and indirectly supports the warfighter. It provides programmed depot maintenance (PDM) and related services directly to the warfighter, and it repairs components for the Supply Maintenance Activity Group (SMAG), which in turn provides them to the warfighter.

Positing a relationship between flying hours and DMAG expenditures requires a series of assumptions. More flying must generate more broken parts. More broken parts must flow from installations to the SMAG. The SMAG must pass more broken parts to the DMAG. The DMAG must perform extra work and this extra work must increase DMAG expenditures. Such a process is more relevant to component repair than to PDM, because PDM does not take place directly in response to recent or current flying, whereas component repair does. For component repair, such a process should involve a series of lags of uncertain length.

EMPIRICAL FINDINGS

We find, across a variety of weapon systems, that although both flying hours and official DMAG repair expenditures for component repair vary considerably month-to-month, there is no consistent, across-weapon-system relationship between the series. We come to the same nonfinding using an accrued expenditure approach we developed that we believe more accurately records how expenditures occur.

We find some suggestion that cargo and tanker aircraft have a closer connection between flying hours and DMAG expenditures than is true for fighters and bombers. This result could be explained by less-robust base-level support of cargo and tanker aircraft. We view this finding as speculative but worthy of further examination.

CONCLUSIONS AND IMPLICATIONS

We expected and found that many DMAG costs (e.g., programmed maintenance, overhead) are unrelated to flying hours. Indeed, we could not find any category of organic DMAG expenditures that is consistently positively correlated with flying hours across multiple weapon systems.

The lack of consistent positive correlation between DMAG expenditures and fleet flying hours argues for an alternative approach to budgeting and pricing. Budgets should reflect a general level of capability maintained in AFMC, regardless of the level of flying hours in a particular year. Prices that operating commands pay for AFMC services should reflect the underlying cost structure in AFMC. If they did, prices would not place nearly as much emphasis on individual repairs as current prices do. Rather, prices would focus on the total costs of maintaining, in AFMC, the level of capability that each operating command wants. The new cost recovery and pricing system recommended by the Air Force Spares Campaign(sometimes called "multi-part pricing") represents a useful move in this direction. Under such an approach, AFMC would receive a budget to pay for a portion of its fixed costs and operating commands would no longer face prices that include these fixed costs. The findings in this report suggest that the Air Force could define fixed AFMC costs to cover a much broader set of costs.

ACKNOWLEDGMENTS

This research was sponsored by Brigadier General Robert Mansfield, U.S. Air Force Directorate of Supply (AF/ILS).

The authors appreciate the assistance of numerous Air Force personnel at Wright-Patterson Air Force Base and Hill Air Force Base in obtaining the H036A expenditure and workload data. Valerie Robertson and Jeff Jones at Hill were particularly helpful. Mike Cerda (AFMC/FMR), Tom Meredith (AF/ILSY), and Michelle Rachie (AF/ILSY) also assisted us. This research was briefed to Edward Koenig, Director of the Aircraft/Missile Support Division of the Air Force Directorate of Supply (AF/ILSY), and his staff on August 17, 2001.

We received constructive reviews of an earlier draft of this research from RAND colleagues Greg Hildebrandt and Ray Pyles. Matt Schonlau provided statistical advice. Mary Wrazen created Figure 5.1. Jeanne Heller edited this document. The authors also thank colleagues Laura Baldwin, Mary Chenoweth, Chris Hanks, Rodger Madison, Gary Massey, C. Robert Roll, and Hy Shulman for their assistance on this research. Randy King, Ginny Mattern, and Robert Steans of the Logistics Management Institute provided helpful input. Earlier versions of this research were briefed at the United States Air Force Academy on November 20, 2000 and to seminars at RAND on May 18, 2001, July 2, 2001, and August 2, 2001. Also, this research was briefed to the American Institute of Astronautics and Aeronautics Economics Technical Committee on January 23, 2002. The insights of seminar participants were appreciated.

Of course, remaining errors are the authors' responsibility.

ABBREVIATIONS

AFCAIG	Air Force Cost Analysis Improvement Group
AFMC	Air Force Materiel Command
AIMD	Accounting and Information Management Division
CINC	Commander-in-Chief
Df	Degrees of Freedom
DLR	Depot-Level Reparable
DMAG	Depot Maintenance Activity Group
FH	Flying Hours
FMR	Directorate of Financial Management and Comptroller, Working Capital Funds Division
FY	Fiscal Year
G&A	General and Administrative
GAO	General Accounting Office
ILS	Deputy Chief of Staff for Installations and Logistics, Directorate of Supply
ILSY	Aircraft/Missile Support Division
JON	Job Order Number

LRU	Line Replaceable Unit
MC	Mission Capability or Mission Capable
MD	Mission Design
MDS	Mission Design Series
MERLIN	Multi-Echelon Resource and Logistics Information Network
MICAP	Mission Capability
MISTR	Management of Items Subject to Repair
MSD	Materiel Support Division
NATO	North Atlantic Treaty Organization
NOR	Net Operating Result
NSIAD	National Security and International Affairs Division
OSD	Office of the Secretary of Defense
P	Probability
PDM	Programmed Depot Maintenance
PPBS	Planning, Programming, and Budgeting System
RCCC	Resource Cost Control Center
REMIS	Reliability and Maintainability Information System
RGC	Repair Group Category
RSP	Readiness Spares Package
SE	Standard Error
SMAG	Supply Maintenance Activity Group

| SS | Sum of Squares |
| WCF | Working Capital Fund |

Chapter One

INTRODUCTION

How do Air Force Materiel Command (AFMC) costs relate to operating command activity levels?

The Planning, Programming, and Budgeting System (PPBS) currently requires operating commands to develop budgets that will cover costs of relevant AFMC support services. The operating commands use the Air Force Cost Analysis Improvement Group (AFCAIG) process to do that. Roughly speaking, this process divides expected expenses into fixed and variable components and develops a budget estimate for each. We are particularly interested in how the methods used to address the variable component, expressed in terms of cost-per-flying-hour factors, relate to the actual costs in AFMC.

In this study, we try to better understand the cost structure of depot support services. How much do these costs change when operating command activity levels change? It is well known that many measures of activity level—like flying hours, sortie length, engine cycles, and landings—can be important cost drivers in particular circumstances. Several may be important at the same time in any particular circumstance. For simplicity, we focus on the single factor that the PPBS process currently emphasizes. Because the AFCAIG process focuses on flying hours as an activity driver, we focus on flying hours as well. It may be useful for future analysis to examine the relevance of other potential cost drivers; however, understanding alternative cost drivers is not critical to the analysis at hand.

We believe the analysis presented here raises questions about how variable costs are reflected in the Air Force's command-based budgeting system. Specifically, we find that the expenditures of AFMC's

Depot Maintenance Activity Group (DMAG) are inconsistently correlated with flying hours across different weapon systems.

DEPOT-LEVEL FIXED COSTS

We hypothesized ex ante that many depot-level costs are unrelated to operating command activities. For example, programmed depot maintenance (PDM) is scheduled years in advance and is unrelated to current operations. The current AFCAIG process recognizes this phenomenon.

The Air Force command-level budgeting process treats unscheduled maintenance costs in the year of execution as though they were driven by operating command flying hours in that year. While there may be evidence that costs from the standpoint of the major commands are related to flying hours, we show that costs actually incurred by AFMC are not related to flying hours in the short run.

For example, materiel procurement has long lead times. Items bought with funds obligated in the year of execution need not be delivered to support flying hours for a year or more. Similarly, items delivered and paid for in the year of execution need not experience a demand in that year.

As we will discuss in more detail below, the organic DMAG has considerable general and administrative (G&A) and overhead costs. By definition, these costs are essentially independent of system activity in any year of execution.

There are also considerable rigidities in the government-employed civilian labor force. See, for instance, Robbert, Gates, and Elliott (1997). Government civilian personnel policy strongly limits AFMC's ability to use temporary labor. Also, the skills required in one maintenance shop are typically too specialized to allow much substitution of available workers between shops to match fluctuations in shops' demands over time.

Even if labor were freely flexible, AFMC is only now systematically attempting to match maintenance shop priorities to priorities in the operating commands. Until a systematic match can be taken for granted, there is no reason to expect a demand generated by an operating command's flying hours to match a maintenance action in AFMC.

What depot-level costs vary with operating command activity levels? The major commands employ operating and support cost factors for Depot-Level Reparables (DLRs) based on calculated costs to them per flying hour. Although these factors are quite helpful in forecasting the DLR costs to the major commands, they are much less successful in predicting how a change in flying hours in any month would affect actual AFMC costs during the months following that month. Our analysis suggests that cost factors, like average cost per flying hour, relevant for long-term purposes, have great difficulty explaining how changes in operating command activity levels affect actual depot-level costs to the Air Force in a particular month.

ROADMAP

Chapter Two explains our analytic approach. Chapter Three provides a stylized overview of the supply chain, and explains the linkages that contribute to any connection between flying hours in the operating commands and expenditures recorded in depot maintenance shops. Chapter Four summarizes our empirical findings. Chapter Five discusses the conclusions' policy implications. The appendices present more detail underlying the analysis.

Chapter Two

RESEARCH APPROACH

Our analysis focuses on the DMAG, where the majority of AFMC logistics support costs are incurred. The DMAG funds all programmed and nonprogrammed maintenance in AFMC. The DMAG buys materiel from the Supply Maintenance Activity Group (SMAG), but unless this materiel is new, the DMAG is responsible for returning it to serviceable status, so much of what the DMAG pays the SMAG simply covers DMAG costs incurred earlier. The only SMAG costs not actually incurred in the DMAG earlier cover purchases of new materiel and administrative costs within the SMAG; as a result, costs in the DMAG drive total costs in the Air Force working capital fund (WCF).

We examine recorded DMAG expenditures, not Air Force Cost Analysis Improvement Group (AFCAIG) cost factors. AFCAIG cost-per-flying-hour factors attempt to capture what the operating commands pay AFMC for services. In this analysis, we are not looking at what the operating commands pay AFMC; this is a paper transfer within the Air Force. (As discussed in Appendix C, however, this paper transfer is a real cost from the perspective of the operating command.) Instead, we are concerned with the actual costs to the Air Force as a whole of supporting the operating commands. AFMC incurs these costs when it pays its labor, buys materiel, pays for utility services, and so on. These are the costs we examine.

We focus on DLR repair expenditures, not PDM. According to the data we obtained, repair represented 42 percent of DMAG expenditures in fiscal year 2000 (FY00).

We focus on specific weapon systems in the operating commands as the drivers of maintenance activities in AFMC. We are concerned with that fraction of repair expenditures that is unambiguously attributed to specific Mission Designs (MDs). In FY00, only about half of organic DMAG repair expenditures was attributed to a specific weapon system. (The rest was attributed to non-system-specific categories like "Engines, Turbines, and Compressors" and "Communication, Detection, and Radio Equipment.")

We use fleet flying hours as our proxy for operating command activity. As noted above, we do this because the AFCAIG process uses this approach as well. We recognize that other cost drivers may be important.

We then use various lags of fleet flying hours to try to find some relationship with repair expenditures.

H036A DATA

Our analysis is based on the DMAG's organic H036A data. These data record monthly accumulated costs in the DMAG by Job Order Number (JON). The Air Force does not attribute actual expenditures to individual JONs. Instead, we infer that these data are based on the following logic: The Air Force can track actual expenditures per month by resource cost control center (RCCC)—typically an individual depot shop. Using standard hours for every maintenance task, it allocates actual civilian labor costs by RCCC to all JONs in progress in that RCCC per month. If this inference is correct, the data we have on costs per JON include distortions where systematic differences persist between actual and standard hours for particular kinds of maintenance actions. But we use these data to construct cost per period at the MD level. Such systematic differences should not bias our analyses of cost relationships between total AFMC costs per month and operating activity levels per month, by MD.

Unfortunately, we have not been able to corroborate our inferences about how data are accumulated in the H036A. The Air Force does not directly use the data in H036A for its own management purposes; it maintains the data solely to report them to the Office of the Secretary of Defense (OSD). We used these data because we were told by several Air Force sources that they provide the best data available for attempting to link MD support costs as closely as possible to actual costs in AFMC. We were able to use the data from H036A to duplicate the DMAG costs that AFMC reported in its *Fiscal Year 2000 Annual Report* on the Air Force WCF. This match gives us an important degree of confidence in the data's integrity.

These data cover all costs relevant to maintenance—direct labor, materiel, overhead, and so on—and distinguish these cost categories. Each record indicates a month, whether a job is completed or is in progress, where the work is occurring, the customer, the MD, and various types of expenditure (e.g., direct civilian labor, materiel, overhead). We have monthly H036A data for fiscal years 1997–2000, inclusive.

Figure 2.1 breaks out FY00 organic DMAG expenditures by materiel, operating overhead, direct civilian labor, and G&A for both programmed and repair work. (Figure 2.1 does not display expenditures

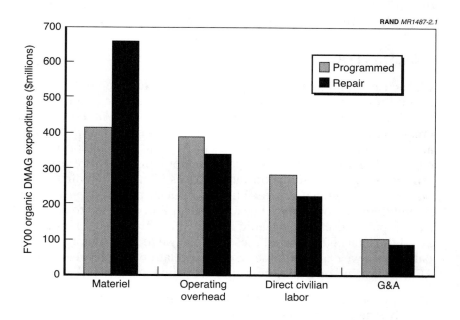

Figure 2.1—FY00 Organic DMAG Expenditures ($ millions)

we categorized as neither programmed nor repair. The largest such "Other" categories are "Software" and "Exchangeables Service Work.")

"Materiel" covers the costs of all DMAG purchases from the SMAG. Note that these are not actual costs to the Air Force as a whole, but transfers from the DMAG to the SMAG. Even if we see a strong relationship between activity levels and materiel costs, we cannot infer a strong relationship between activity levels and actual costs to the Air Force in the year of execution.

Operating overhead and G&A are clearly fixed costs in the year of execution. Note that operating overhead exceeds direct civilian labor costs. Direct civilian labor represents a distinct minority of organic DMAG expenditures.

Direct civilian labor costs are the most likely cost category shown to reflect a direct relationship between activity levels and actual costs to the Air Force as a whole. But, as noted above, they too may well display only a limited relationship in the year of execution.

ANALYSIS FOCI

We focused our analysis in several ways. We examined only organic expenditures attributed to MDs. We looked at organic expenditures, because over the period that our data cover, the Air Force gave greater emphasis to tightening order-and-ship times between operating commands and repair facilities for organic than for contract facilities. Similarly, the Air Force gave greater emphasis to prioritizing repairs in organic than in contract facilities. Contract repair also reportedly experienced greater budget-induced turbulence over this period than organic repair, because the Air Force tended to cut contract repair more than organic repair when funds were short. To avoid the likely effects of all these factors, we focused on organic repair, where we expected to see a cleaner relationship between actual AFMC costs and operating command activity levels.

More prosaically, the organic H036A data are more detailed and descriptive than contract H036A data.

Within organic DMAG expenditures, we looked at total repair costs [or what H036A calls "Management of Items Subject to Repair

(MISTR)" expenses], which is what the AFCAIG process emphasizes in its treatment of variable costs.

As we will discuss, we took two perspectives on DMAG expenditures. The first uses expenditures associated with completed JONs during a month as the unit of observation. This "official" version of the data accords, as noted above, with the FY00 AFMC annual report. The second perspective analyzed "accrued" expenditures using monthly first differences of work-in-progress expenditure totals. Such first differences measure changes in the value accumulated against JONs.

We restricted ourselves to MDs for which we had monthly fleet flying hours as well as H036A expenditure data. The flying hours data came from the Air Force's Multi-Echelon Resource and Logistics Information Network (MERLIN) system, which obtains the data from the Reliability and Maintainability Information System (REMIS). We analyzed the B-1, B-52, C-130, C-135, C-141, C-5, F-15, and F-16.

We have flying hours at the Mission Design Series (MDS) level (e.g., F-15Cs), but the H036A data tend to be only by MD (e.g., F-15s).

ESTIMATION PROCEDURE

Our basic estimation procedure was to estimate the parameters of the equation

$$Expend(t) = a + \sum_{i=0}^{i=12} b(i)FH(t-i) + \varepsilon,$$

where Expend(t) is DMAG repair expenditures in month t in support of this MD and FH(t − i) is total Air Force flying hours for this MD i months prior to month t. We regressed, for a given MD, monthly expenditure data on current-month and 12 monthly lags of flying hours. We assume that the residual term, ε, is well behaved with zero expectation, independence between observations, and a fixed variance.

If increasing flying hours increases expenditures, we expect the sum of the 13 b(i) estimates to be positive.

One could consider other potential cost drivers. The number of observations (47 or 48, depending on whether we use accrued or official expenditure data) limits the number of independent variables one can consider. Adding additional variables will probably require use of a simpler lag structure. The data force us to accept such analytic trade-offs. An alternative would be to add years to the database. But Air Force logistics policy has been dynamic enough since 1991 to make additional years suspect; with each additional year of data, the argument that we are measuring a stable underlying structure—one of the assumptions of linear regression analysis—becomes harder to sustain. Also, H036A data were not readily available to us for years before FY97.

Chapter Three

LINKING DMAG EXPENDITURES TO OPERATING COMMAND FLYING HOURS

Figure 3.1 roughly describes the DMAG's position in the Air Force.[1] The DMAG directly and indirectly supports the warfighter. It provides PDM directly to the warfighter at the flight line. It provides component repair to the SMAG, which in turn provides serviceable parts as needed to the flight line. The SMAG covers all materiel in the Materiel Support Division (MSD) and base supply.

Our econometric approach measures the extent of a relationship between the wing flight line, where flying hours are measured, and organic component repair shops in the DMAG.

DO MORE FLYING HOURS INCREASE DMAG EXPENDITURES?

A series of conditions must hold true for increased flying hours to increase DMAG expenditures.

First, more flying must generate more parts needing servicing.[2] We know that the Air Force expects flying hours to drive failures for only

---

[1] Extensive flows of services, serviceable and unserviceable materiel, money, and information accompany the processes in the supply chain. To keep things as simple as possible, we depict only flows of services and serviceables from sources of supply and repair to users.

[2] There is an extensive literature on the presence of uncertainty in the generation of demand on the supply chain from flying hours. See, for example, Brown (1956),

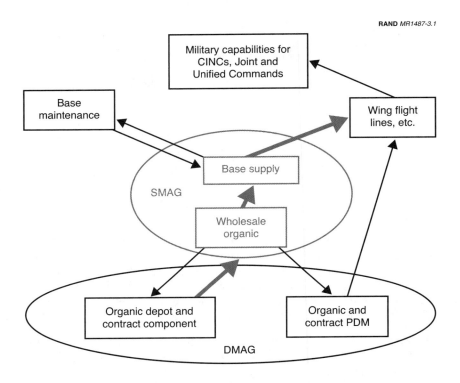

Figure 3.1—The DMAG's Position in the Air Force

a portion of its aircraft-related inventory. In some cases, historical data indicate that another driver—perhaps sorties, cycles, or operational hours—is a better predictor of failures. Failure-induced demand for many items is related to the size of the inventory, but not to any specific activity level. And for many items, no empirical relationship has ever performed well enough to predict demand. Items falling into all of these categories drive workload in the DMAG. Flying hours are relevant only to those items for which (1) flying hours are the primary driver or (2) the primary driver is correlated with flying hours. Any relationship that we capture will be better de-

---

Hodges (1985), Crawford (1988), Cohen, Abell, and Lippiatt (1991), Adams, Abell, and Isaacson (1993), Pyles and Shulman (1995), and Feinberg et al. (2001).

fined as items with a demand driven by flying hours become more dominant in DMAG's workload.

Second, these failure-induced demands for additional parts must pass from base maintenance into the SMAG. That is, items that require servicing must generate a demand in base supply. To the extent that base maintenance relies on cannibalized line replaceable units (LRUs) or aircraft to fill a demand on the flight line, it can delay the time at which a demand is generated in base supply. Within the SMAG, base supply must then pass an effective demand on to wholesale supply. If base supply draws down a stock level, wholesale supply can decide not to fill it immediately and thereby avoid passing the demand along.

In the last few years, operating commands have complained increasingly that the SMAG was slow to fill depleted base-level stocks because doing so did not generate sales for the SMAG. To the extent that this is true, the link between a demand generated at the flight line and a demand on the DMAG is further weakened.

Third, the SMAG must pass additional demands for replacement parts on to the DMAG. To do so, it must present a requisition and commit resources to cover the cost of servicing the requisition. Because Air Force policy on stock leveling allows the SMAG to receive more requisitions than it can pay for, it will tend to pass on requisitions first for items it will have the least difficulty selling to the operating commands. That is, the SMAG may delay placing a demand on the DMAG to avoid expenditures that would degrade its financial performance. In the extreme, if the SMAG buys too many parts from the DMAG that it cannot sell, it may not have the financial capacity to buy anything from the DMAG for some period, even if the DMAG has resources available to commit to repair.

Fourth, the DMAG must perform additional work. To do so, it must (1) have all the parts required to perform the maintenance, including an unserviceable carcass, and (2) give the repair enough priority to induct the item in question. Without the basic repair capacity and priority, a specific induction can wait, loosening the link between the operating command and depot-level maintenance costs.

Finally, this extra work must increase DMAG expenditures. This assumption is not trivial in that we believe the DMAG has a large number of fixed costs that do not increase with workload. In principle, expenditures occur immediately when the DMAG commits civilian labor to a task and pays for materiel and other inputs to support the repair. As noted above, it is not clear how closely H036A links any repair action to recorded expenditures. More broadly, any delays between actual and recorded expenditures in other data systems that the Air Force uses to manage depot maintenance will dampen the link between an operating command and recorded costs in ways that we cannot observe directly.

These concerns make each of the links necessary to connect flying hours and DMAG expenditures problematic. Some failures lead to demands that pass through the supply chain extremely quickly. Other failures generate demands that languish in the supply chain and generate work in a DMAG shop long after, if ever.

A NOISY LINK

In sum, we expect the link between flying hours and recorded DMAG expenditures to be very noisy.

The initial failure process is noisy. Variance-to-mean ratios of 5 or more have been observed for some parts. See, for instance, Crawford (1988). With low probabilities of failure, there can be extreme variability from year to year in the number of failures generated by a constant flying hour program. This variability alone could prevent an analytic method that is suitable for developing long-term cost factor averages from yielding useful information for any particular future year.

Many demands upon the DMAG result not from flying hours but from other measures of operating command activity levels. These exogenous demands are also likely to be highly stochastic, injecting more variation in total DMAG demands that has no direct connection to flying hours at all.

Initial demands pass through the supply chain in idiosyncratic ways. If the Air Force had optimized the supply chain and made decisions at each level to support optimal outcomes, the link between operating command demands and DMAG costs might tighten. Perhaps a better system would attempt to (1) match repair actions to actual demands as closely as possible to avoid unnecessary costs, and (2) standardize the pipeline lengths at as low a level as possible to reduce demand for pipeline inventories.

The distributed lag structure discussed at the end of the preceding chapter could be estimated to link demand to DMAG cost, reflecting the unavoidable delays in the pipeline. However, it could be that the basic lag structure for the logistics pipeline is stochastic and may not be stable over significant periods of time, adding noise to any attempt to identify the underlying structure.

High variance-to-mean ratios in initial demands, additional variability added from demands not driven by flying hours, and a stochastic lag structure lead to significant variability in demands placed upon the DMAG in any period. These challenges would suggest that, even if both the underlying stochastic structure of each element in this system and flying hours remained constant, the realized demands on the DMAG would vary substantially over time.

In contrast, we know that DMAG overhead and G&A costs are fairly stable. Direct civilian labor is fairly stable. So it is highly unlikely that performed workload in DMAG shops will vary enough to accommodate the demand variability implied by the above factors. The DMAG will be managed to absorb this demand uncertainty in an effort to keep its own direct civilian labor fairly steadily employed and to limit variability in demand for materiel inputs from the SMAG.

How will such accommodation affect the link between demands in the operating commands and reported costs in the DMAG? Any DMAG activity will accommodate such variation by working down standing backlog or standing idle when shop capacity exceeds current demand or allowing backlog to accumulate when current demand exceeds shop capacity. Changing the level of backlog simply moves the time when the DMAG services any particular demand from an operating command. Such accommodation is likely to introduce additional discretion that further dilutes any relationship between flying hours and DMAG costs.

Chapter Four

EMPIRICAL FINDINGS

Our analysis produced three basic empirical findings.

- First, flying hours and DMAG organic repair expenditures varied significantly over time for all weapon systems examined. This variation provided the basis for the remaining empirical analysis.

- Second, we found no strong patterns of results that suggest a uniform relationship between flying hours and DMAG organic repair expenditures across weapon systems. To the extent that any relationships exist, they differ substantially across systems.

- Third, the data suggest that increased flying hours may increase subsequent DMAG expenditures associated with cargo-type aircraft while reducing subsequent DMAG expenditures associated with fighters and bombers.

This chapter discusses each of these findings in turn.

VARIABILITY IN FLYING HOURS AND DMAG ORGANIC REPAIR EXPENDITURES

Both flying hours and DMAG organic repair expenditures vary considerably month-to-month across all weapon systems studied. Figure 4.1 illustrates this pattern for the C-135. The early 1999 spike in C-135 flight hours, for example, is related to North Atlantic Treaty Organization (NATO) operations in the Balkans.

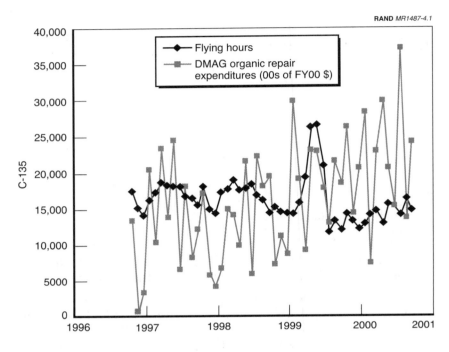

Figure 4.1—C-135 Flying Hours and DMAG Organic Repair Expenditures

From a purely analytic point of view, the degree of variability in both data series is a good thing. Without it, we would be unable to measure any correlation between recorded DMAG organic expenditures and operating command activity levels. We worried initially that careful management in the operating commands and in the DMAG shops would prevent the variation we needed to proceed. So this result is very important.

GENERAL RELATIONSHIPS BETWEEN FLYING HOURS AND DMAG ORGANIC REPAIR EXPENDITURES

We observed no straightforward relationship between the DMAG's organic repair expenditure and flying hour series that is consistent across weapon systems.

Viewing the same C-135 data a different way, Figure 4.2 plots quarterly C-135 flying hours against C-135 DMAG organic repair expenditures covering FY97–FY00. There is no obvious relationship.

The figure looks essentially the same if one looks at a quarterly lag of flying hours as the independent variable.

Figure 4.3 uses the same C-135 data, but on a monthly basis. Again, there is no clear relationship between flying hours and DMAG organic repair spending.

Formalizing the intuition of Figures 4.1–4.3, Table 4.1 presents the results of regressing C-135 DMAG organic repair expenditures on current-month plus 12 monthly lags of C-135 fleet flying hours.

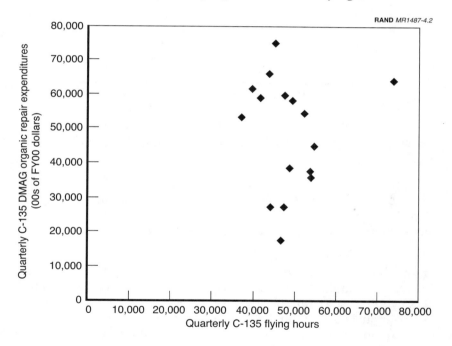

Figure 4.2—Quarterly C-135 Flying Hours and DMAG Organic Repair Expenditures

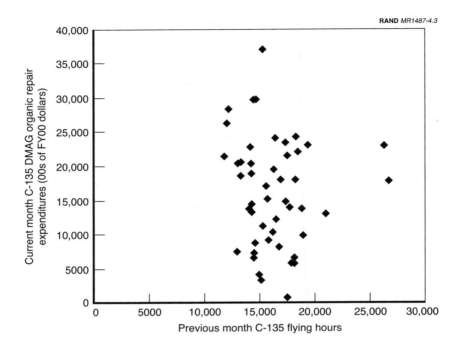

Figure 4.3—Monthly C-135 Flying Hours and DMAG Organic Repair Expenditures

None of the monthly fleet flying hour coefficient estimates is significant in Table 4.1. There is no clear evidence of C-135 fleet flying hours influencing organic DMAG C-135 repair expenditures.

One possible problem with Table 4.1 is that the various lags of monthly flying hours are highly correlated with one another, so the estimation should be run with fewer independent variables. To test this hypothesis, we ran the stepwise regression procedure where it is endogenously determined which independent variables should be used in the model estimation. Table 4.2 gives the result of the stepwise estimation on these data.

The stepwise procedure resulted in a much more parsimonious regression structure in that it had many fewer independent variables. But, reiterating Table 4.1's results, there is no immediate evidence

Table 4.1

**Monthly Organic C-135 DMAG Repair Expenditures
Regressed on Monthly Fleet Flying Hours, FY97–FY00**

Observations		48		
R-squared		0.25		
F		0.89		
Pr > F		0.57		
		Df	SS	
Regression		13	7.6E12	
Residual		34	22.4E12	
Total		47	30.0E12	
	Coefficient	SE	T-Statistic	P-Value
Intercept	4460558	2194804	2.03	0.05
Current month	–6.35	60.98	–0.10	0.92
FH-1[a]	–32.13	78.72	–0.41	0.69
FH-2	39.99	77.63	0.52	0.61
FH-3	–106.24	80.86	–1.31	0.20
FH-4	18.25	80.81	0.23	0.82
FH-5	63.22	81.58	0.77	0.44
FH-6	–152.17	85.53	–1.78	0.08
FH-7	82.67	83.40	0.99	0.33
FH-8	–25.57	83.81	–0.31	0.76
FH-9	–52.16	84.94	–0.61	0.54
FH-10	–91.09	81.40	–1.12	0.27
FH-11	159.26	81.66	1.95	0.06
FH-12	–70.92	65.71	–1.08	0.29

[a]FH-1 refers to fleet flying hours lagged one month, FH-2 is fleet flying hours lagged two months, and so forth.

that C-135 fleet flying hours have a marked impact on DMAG organic C-135 repair expenditures.

RELATIONSHIPS BASED ON ACCRUED EXPENDITURES

We were surprised at the month-to-month variability of the DMAG expenditure data in the preceding figures. One explanation, we believe, is that the DMAG does not systematically use accrual accounting. Instead, expenditures are booked sporadically, i.e., when a JON is closed out. We undertook a work-in-progress monthly first-

Table 4.2

Monthly Organic C-135 DMAG Repair Expenditures Regressed Stepwise on Monthly Fleet Flying Hours, FY97–FY00

	Coefficient	SE	T-Statistic	P-Value
Observations		48		
R-squared		0.11		
F		2.84		
Pr > F		0.07		
		Df		SS
Regression		2		3.4E12
Residual		45		26.6E12
Total		47		30.0E12
Intercept	1108024	726696	1.52	0.13
FH-10	−90.05	50.92	−1.77	0.08
FH-11	121.06	51.32	2.36	0.02

difference accrual adjustment for this phenomenon.   This adjustment implies a divergence from the data in the AFMC FY00 annual report.   But we believe it more nearly captures the actual expenditures incurred per month in DMAG shops.

To illustrate how our accrual adjustment process worked, Table 4.3 shows a specific JON (associated with C-130 work at Warner Robins Air Force Base) and its progression of total organic expenditures between September 1999 and June 2000.   This example illustrates how the official tabulation method creates an artificially clustered and lagged portrayal of expenditures relative to the accrued methodology.

Figure 4.4 shows the C-135 organic repair expenditures, according to the official H036A data, compared to the pattern using our accrual adjustment. The accrued repair expenditures show considerably less month-to-month variability, as one would expect.

It turns out, however, that none of our results concerning the lack of a consistent connection between flying hours and DMAG organic repair expenditures is changed based on whether one uses official versus accrued expenditure data. Table 4.4 gives the C-135 stepwise re-

Table 4.3

Official and Accrued Expenditures of JON 00081B417

Month	Cumulative Expenditures	Official	Accrued
September 1999	28872	0	28872
October 1999	332774	0	303902
November 1999	570108	0	237334
December 1999	888240	0	318132
January 2000	1185753	0	297513
February 2000	1185753	0	0
March 2000	1510393	0	324640
April 2000	1514941	0	4548
May 2000	1524771	0	9830
June 2000	1524775	1524775	4

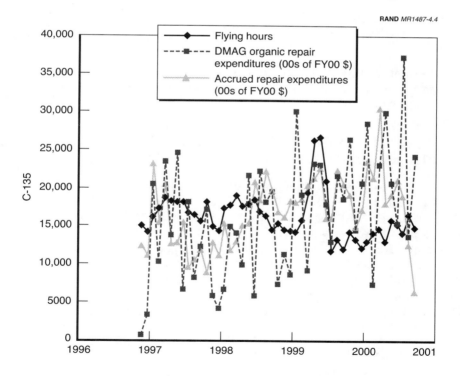

RAND *MR1487-4.4*

Figure 4.4—C-135 Flying Hours and DMAG Accrued Repair Expenditures

Table 4.4

**Monthly Organic C-135 DMAG Accrued Repair
Expenditures Regressed Stepwise on Monthly Fleet
Flying Hours, FY97–FY00**

		Df	SS
Observations		47	
R-squared		0.24	
F		6.79	
Pr > F		0.00	
Regression		2	2.3E12
Residual		44	7.6E12
Total		46	10.0E12

	Coefficient	SE	T-Statistic	P-Value
Intercept	21657	467958	0.05	0.96
FH-10	65.12	21.82	2.98	0.00
FH-12	34.78	21.84	1.59	0.12

gression results where the dependent variable is accrued repair expenditures. In Appendix A, we present similar analyses for other weapon systems.

POTENTIALLY DIFFERENT PATTERNS FOR CARGO AIRCRAFT AND FOR FIGHTERS AND BOMBERS

An interesting pattern emerges when we look at the implications of the regressions estimated in a slightly different light.[1] Suppose we use these equations to ask how much accrued DMAG organic repair expenditures would rise if flying hours rose by one hour. Table 4.5 summarizes the results of asking this question in three different ways. It asks how an extra flying hour would affect accrued DMAG expenditures over the 13-month period we have examined—the month concurrent with the increase in flying and the 12 following months. Results differ depending on which form of the model we use to answer this question.

---

[1]The analysis that follows benefits from discussions with Ray Pyles.

Table 4.5

**How One Extra Flying Hour Affects Accrued DMAG
Organic Repair Expenditures Over 13 Months**

System	Full Regression	Stepwise Regression	Constrained Regression
C-130	$1401	$683	$689
C-135	$76	$100**	$39
C-141	$194**	$187**	$182**
C-5	$1189	$1943*	$1807
B-1	−$693	−$632*	−$1027**
B-52	−$2879**	−$2342**	−$2132**
F-15	−$2089	–	−$1716*
F-16	−$1710**	−$1414**	−$1560**

NOTE: * denotes a sum that is statistically significant at the 95% confidence level; ** denotes 99% significance.

Column 1 lists the eight weapon systems studied. Column 2 shows the implication of applying a full regression, like that reported in Table 4.1. Column 3 shows the implication of applying a stepwise regression, like that reported in Tables 4.2 and 4.4. Column 4 shows the implication of applying a full regression, but constraining all the coefficients on flying hours to be equal.[2] The asterisks show when the sums of the relevant coefficient estimates are statistically significantly different from zero. Tables A.14 and A.15 provide more detail on the estimations underlying Table 4.5.

The answers for each weapon system are roughly the same across models, but the statistical significance of the findings varies considerably. That said, it is hard to dismiss a systematic difference in results for cargo aircraft and for fighters and bombers. These results suggest that an additional flying hour is likely to increase DMAG expenditures relevant to cargo aircraft, but to decrease expenditures relevant to fighters and bombers. This was a totally unexpected outcome.

Further work is required to determine whether this result is real and, if it is, what causes it. Our colleague, Ray Pyles, offers the following

[2]Greg Hildebrandt suggested this type of model to test the robustness of our results in the face of multicollinearity. As the results in Table 4.5 demonstrate, the findings are quite robust across the three functional forms.

hypothesis as one worth exploring further: Cargo and other support aircraft tend to receive less ample spares levels than do bombers and fighters because

- their aircraft availability goals are set lower,

- their de facto operating configuration (worldwide dispersed operations) is ignored, thereby underestimating required safety levels, and

- the C-5 and C-141 war reserve readiness spares package (RSP) computations assume operations from one base instead of the dispersed squadron deployment assumed for combat aircraft.

As a result, a support aircraft DLR demand surge (perhaps caused by a higher operational tempo) will run through the available spares more quickly than a proportional surge in combat aircraft DLR demands. Any additional demands will cause "holes" in aircraft, called mission capability (MICAPs) failures, which receive higher priority in both depot repair and transportation functions than other demands. If demands increase in response to a forcewide operational tempo surge, the combat aircraft will experience relatively fewer MICAPs than support aircraft and more of the available DMAG repair capacity will be devoted to the cargo and other support aircraft. This workload emphasis shifts because the SMAG and the DMAG work with fixed overall budgets, and so can only reallocate workload to minimize the worst effects of a demand surge.

Table 4.6 shows that the flying hours for the eight weapon systems studied here do, in fact, tend to move together. Cross correlations are significant between almost all of these systems and are often quite high. This pattern lends credence to the Pyles hypothesis described above. We urge further research on the potential differential treatment of the support and combat fleets in the depot system.

SUMMARY

The current method the Air Force uses to build budgets for the operating commands to pay for DMAG services leads one to believe that actual DMAG expenditures are proportional to flying hours and that flying hours can explain the preponderance of DMAG expenditures in any year. The results reported here fail to support this view

Table 4.6

Correlations of Monthly Fleet Flying Hours, FY97–FY00

System	B-1	B-52	C-130	C-135	C-141	C-5	F-15
B-52	0.52**						
C-130	0.58**	0.47**					
C-135	0.36*	0.41**	0.66**				
C-141	0.09	0.24	0.31*	0.46**			
C-5	0.19	0.26	0.40**	0.43**	0.49**		
F-15	0.67**	0.54**	0.76**	0.76**	0.37*	0.31*	
F-16	0.70**	0.55**	0.84**	0.75**	0.38**	0.42**	0.91**

NOTE: * denotes a sum that is statistically significant at the 95% confidence level; ** denotes 99% significance.

in two different ways. First, the specific relationships we find between DMAG organic repair expenditures and flying hours are quite idiosyncratic across weapon systems. Second, it may be that increased flying hours increase DMAG expenditures for aircraft that the Air Force gives relatively few spare parts and decrease DMAG expenditures for aircraft that the Air Force favors with more spare parts. Whether or not this hypothesis turns out to be true on closer examination, the current budgeting system clearly assumes a relationship between flying hours and actual costs in AFMC that is not found if one examines the relationship between actual DMAG costs and flying hours in the months preceding these costs.

Chapter Five

CONCLUSIONS AND IMPLICATIONS

We expected and found that many DMAG costs are unrelated to fly-ing hours. Programmed maintenance, by definition, is independent of current operating command activity levels. A sizable fraction of DMAG expenditures goes to output-invariant costs like G&A and overhead. Also, government civilian worker regulations tend to make labor costs hard to quickly reduce. Indeed, analysis by Wallace, Kem, and Nelson (1999) suggested that 80 percent of work-ing capital fund costs are fixed with respect to the amount of depot-level reparable sales.

Our multisystem data analysis found a variety of relationship pat-terns between flying hours and organic DMAG expenditures, some in the intuitive positive direction but others in the counterintuitive negative direction. These patterns are not consistent across aircraft. Further, the estimated lag structures vary considerably, both in du-ration and magnitude.

Figure 5.1 provides a conceptual portrayal of what we think is occur-ring. The rows show successive activities in the supply chain, from initial activity at the flight line to the depot maintenance shop. Within a row, time moves from left to right; the Xs show events at each stage in the supply chain that can be traced back to the initial flight-line activity.

At the top of Figure 5.1, we show flight-line activity, i.e., flying air-craft. Flying aircraft probably causes some removals at base, al-though Bachman and Kruse (1994) report only low-to-moderate cor-relation between flying hours and non-overhaul demand across 50 aviation systems.

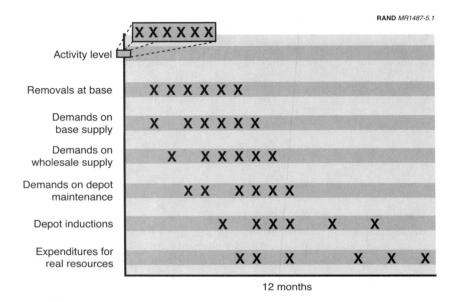

Figure 5.1—A Conceptualization of the Activity-to-Depot Process

Removals at base cause demands on base supply, but perhaps with some lags. Similarly, demands on base supply eventually translate into demands on wholesale supply, but not instantaneously. Managers in base and wholesale supply make a variety of discretionary decisions that serve to diffuse the relationship between demands at the flight line and depot-level activity.

Wholesale supply demands eventually translate into demands on depot maintenance but, again, the process may be lagged based on inventory status, the financial condition of the SMAG, and other factors.

Depot maintenance demands do not instantaneously translate into depot inductions because the depot system might have various backlogs it is managing. Only when work actually occurs do we see depot-level expenditures recorded in H036A. As depicted in Figure 5.1, these expenditures may considerably lag, by uncertain length, the activity that ultimately generated the expenditures.

We are not suggesting flying hours are irrelevant. If the flying hour program changes, we would expect total demand on depot shops to increase. However, (1) it will take a long time, (2) the exact effect in any time period will be uncertain, (3) increased demand will not lead to proportionally increased expenditures even in the long run, and (4) until depot capacity actually changes, increased demand is more likely to increase the backlog than to increase expenditures in the depot.

AN ALTERNATIVE APPROACH TO BUDGETING

Our empirical findings are consistent with an alternative approach to budgeting. This analysis suggests, and other analyses support the view, that the DMAG has significant fixed costs that we would not expect to change with any measure of activity level in the operating commands. The current AFCAIG process recognizes that PDM costs should be viewed in this way.

We believe that many AFMC costs considered variable in the major command budgeting process are also, in all probability, fixed in any year of execution. For example, G&A and overhead costs account for large portions of DMAG component repair costs but are unlikely to respond much to changes in repair workload in the depot.

Our empirical results suggest that even costs that many would link directly to component repair costs, like the costs of direct civilian labor, do not vary proportionately with operating command activity levels in the months leading up to the month when AFMC incurred these costs. In all likelihood, DMAG costs this year depend on past decisions about capacity that affect labor costs today. Tentative results in the previous chapter also raise the possibility that past investment decisions about spares parts may affect DMAG costs today in unexpected ways. Both labor and materiel costs today may be driven more by past investment decisions than by current activity levels. These results suggest that the Air Force should not budget for costs associated with direct civilian labor or materiel costs in the DMAG shops by assuming that they are related to flying hours that occurred in the year leading up to the month in which the DMAG incurred these costs. The logic offered here about uncertainty, lags, discretionary action in the supply chain, and workload smoothing in the DMAG shops helps us understand why no such correlation need be present.

If actual current costs in the DMAG do not depend much on current activity levels in the operating commands, budgeting for these costs is more likely to succeed if it addresses the factors that do drive DMAG costs. Suppose current costs in the DMAG depend more on current depot repair capacity than on current activity levels in the operating commands. The Air Force, in effect, chooses a level of component repair capability in AFMC each year and programs re-sources to provide that capability. In fact, to be successful, ongoing efforts to implement agile combat support and an expeditionary Air Force must focus on proactively choosing a flexible maintenance ca-pability that can meet future uncertain demands when they arise. Total flying hours in any year is only one factor relevant to the design and sizing of such a capability. There is no reason to expect that ag-ile depot-level support for expeditionary forces should display a cost structure that is proportional to flying hours. Once a robust, flexible depot repair capability is in place, the variable cost of servicing indi-vidual flying hours is likely to be small.

We have looked directly only at budgeting in this analysis. But if the empirical results presented here hold up to additional investigation, this budgeting analysis raises questions about WCF pricing as well. The literature on optimal internal transfer prices is clear that prices should reflect the decisions they are designed to inform. If they in-form investment decisions, they should reflect all future costs of in-vestment; if they inform the ongoing use of existing assets, they should reflect only marginal costs associated with marginal use of these assets. Fixed or sunk costs that do not vary with output levels are irrelevant to prices that inform decisions about output. See Baldwin and Gotz (1998).

In this case, WCF prices should reflect information on how the costs of WCF activities, including DMAG shops, vary in response to changes in the operating commands' demand for the services pro-vided by these activities. Our analysis suggests that DMAG costs in any month may vary little in response to changes in the operating commands' demand in the year leading up to that month. The costs move around within the DMAG as the DMAG shifts its attention from one class of repair activities to another. But if total capacity utilization does not change much in response to a surge in demand, marginal costs are likely to be quite low. This finding raises questions about the structure of WCF prices. If AFMC can better budget for funds to cover significant DMAG costs than can the operating commands, perhaps AFMC should take this responsibility and remove relevant funds from the WCF altogether.

This observation accords with skepticism about the appropriateness of current WCF pricing policies expressed in Camm and Shulman (1993), Baldwin and Gotz (1998), Keating and Gates (1999), and Brauner et al. (2000).

The new approach to cost recovery and pricing recommended by the Air Force Spares Campaign represents a useful step toward a pricing structure consistent with the findings reported here. Under this proposal, AFMC would receive a budget to pay for all MSD costs not driven by operating command activity levels. Based on the prevailing wisdom about what AFMC costs are variable, the new cost recovery and pricing proposal suggests that WCF prices would include only direct labor and materiel costs. Our empirical findings suggest that these costs are not proportional to flying hours either, at least not within a specific year. Our findings suggest that even a larger portion of AFMC costs might be removed from the WCF and be budgeted for directly by AFMC. Further analysis of the existing cost structure in AFMC is needed to determine what variable costs an operating command actually imposes on AFMC when it demands a specific service. It is quite possible that operating commands impose costs on AFMC mainly by demanding capacity, not individual repairs. To the extent that this is true, it might be appropriate to orient prices to annual fees for capacity rather than individual repair transactions.

IMPROVED SUPPLY CHAIN INTEGRATION

These thoughts about budgeting and pricing focus on the Air Force supply chain as it currently operates. At any time, budgeting and pricing methods should be compatible with the supply chain as it currently operates. But the supply chain can also improve over time. One way to react to the empirical results reported here is to conclude that the supply chain is not working as well as it could. Certainly, Figure 5.1 does not portray a well-integrated supply chain. If it turns out that indeed an increase in flying hours systematically shifts DMAG focus from combat to support aircraft, the constraints in the supply chain that cause this behavior should be removed.

If the supply chain were better integrated, DMAG expenditures would be better aligned with current activity levels in the operating commands. Once alignment improved, Air Force budgeting and pricing processes should recognize the change and adjust to reflect an appropriate link between budgets and prices on the one hand and activity levels in the operating commands on the other.

DMAG EXPENDITURES—FLEET FLYING HOUR REGRESSIONS

In Chapter Four, we found a limited connection between C-135 fleet flying hours and C-135 DMAG organic repair expenditures. In this appendix, we develop parallel analyses for seven other MDs: the B-1, B-52, C-130, C-141, C-5, F-15, and F-16. Our broad-brush finding is that there is no evidence of a consistent, cross-system, positive flying hour link to DMAG organic repair expenditures.

We first replicate Tables 4.1 and 4.2 for other systems and assess the relationship between official (consistent with the FY00 AFMC annual report) DMAG organic repair expenditures and fleet flight hours. We then replicate and expand upon Tables 4.4 and 4.5 using accrued, rather than official, repair expenditure data.

OTHER SYSTEMS' LINK BETWEEN OFFICIAL DMAG REPAIR EXPENDITURES AND FLEET FLYING HOURS

Table A.1 shows the results of regressing official organic B-1 DMAG expenditures on contemporaneous and 12 monthly lags of B-1 fleet flying hours. Only the current-month flying hour coefficient is positive and significant. A number of other months have negative point estimates.

As in the C-135 case, we accompany our full regression results with results of stepwise regression. Stepwise regression endogenously chooses which independent variables to include. The resultant regression is typically more parsimonious, and arguably more appropriate, than the full regression.

Table A.1

**Monthly Organic B-1 DMAG Repair Expenditures Regressed on
Monthly Fleet Flying Hours, FY97–FY00**

Observations	48			
R-squared	0.35			
F	1.43			
Pr > F	0.20			

	Df	SS		
Regression	13	1.4E12		
Residual	34	2.6E12		
Total	47	4.0E12		

	Coefficient	SE	T-Statistic	P-Value
Intercept	578019	856755	0.67	0.50
Current month	558.51	203.84	2.74	0.01
FH-1	113.05	191.11	0.59	0.56
FH-2	171.83	166.43	1.03	0.31
FH-3	0.58	165.74	0.00	1.00
FH-4	44.12	171.86	0.26	0.80
FH-5	−148.57	170.49	−0.87	0.39
FH-6	21.71	167.36	0.13	0.90
FH-7	−146.23	168.16	−0.87	0.39
FH-8	−10.08	170.32	−0.06	0.95
FH-9	−191.97	166.76	−1.15	0.26
FH-10	−29.49	162.55	−0.18	0.86
FH-11	−78.57	166.11	−0.47	0.64
FH-12	−330.48	174.99	−1.89	0.07

In Table A.2, we show the stepwise results for the B-1 case. Again, the current-month coefficient estimate is positive and significant. However, the negative coefficient estimates for months 7, 9, and 12 imply that increased flying hours eventually have a negative effect on DMAG organic repair expenditures, netting out the positive effect of the current-month coefficient with the negatives in months 7, 9, and 12.

We conclude that there is little evidence of a persistent positive impact of B-1 fleet flying hours on official B-1 DMAG organic repair expenditures.

Table A.2

Monthly Organic B-1 DMAG Repair Expenditures Regressed Stepwise on Monthly Fleet Flying Hours, FY97–FY00

Observations		48		
R-squared		0.29		
F		4.49		
Pr > F		0.00		
		Df	SS	
Regression		4	1.2E12	
Residual		43	2.9E12	
Total		47	4.0E12	
	Coefficient	SE	T-Statistic	P-Value
Intercept	801755	461596	1.74	0.09
Current month	550.03	173.19	3.18	0.00
FH-7	−195.70	131.45	−1.49	0.14
FH-9	−202.85	134.76	−1.51	0.14
FH-12	−294.35	148.96	−1.98	0.05

Table A.3 presents results for the B-52. No monthly lag of fleet flying hours has a significant effect.

Table A.4 presents stepwise results for the B-52. All three chosen monthly lags have negative fleet flying hour coefficient estimates.

The B-52 shows no evidence of fleet flying hours increasing official DMAG organic repair expenditures.

Table A.5 presents the now-familiar pattern of no statistically significant flying hour coefficients, this time for the C-130.

The stepwise procedure (Table A.6) reduced the number of C-130 independent variables to only the five-month lag. The five-month lag coefficient estimate is positive but statistically insignificant.

Table A.7 presents regression results for the C-141. The eight-month lag has the lone statistically significant coefficient estimate and it is negative.

Table A.3

Monthly Organic B-52 DMAG Repair Expenditures Regressed on Monthly Fleet Flying Hours, FY97–FY00

	Observations	48		
	R-squared	0.40		
	F	1.75		
	Pr > F	0.10		
		Df	SS	
	Regression	13	4.0E12	
	Residual	34	5.9E12	
	Total	47	9.9E12	

	Coefficient	SE	T-Statistic	P-Value
Intercept	5210804	1286554	4.05	0.00
Current month	−110.58	278.43	−0.40	0.69
FH-1	−248.61	274.56	−0.91	0.37
FH-2	192.33	273.11	0.70	0.49
FH-3	−250.37	272.94	−0.92	0.37
FH-4	−425.17	256.59	−1.66	0.11
FH-5	−489.28	265.98	−1.84	0.07
FH-6	−193.82	250.38	−0.77	0.44
FH-7	−427.85	263.64	−1.62	0.11
FH-8	86.07	260.18	0.33	0.74
FH-9	187.41	259.38	0.72	0.47
FH-10	−198.27	258.76	−0.77	0.45
FH-11	−287.77	258.21	−1.11	0.27
FH-12	−242.48	272.29	−0.89	0.38

Curiously, the C-141 stepwise procedure (Table A.8) reduces the result to just a six-month lag. Its coefficient estimate is statistically significant and positive but fairly small.

We had to handle the C-5 regression differently. Figure A.1 illustrates the problem: recorded C-5 DMAG organic repair expenditures plummeted in early to mid calendar year 2000. We believe this data problem relates to the relocation of C-5 organic repair work from the closing Kelly Air Force Base to Warner Robins Air Force Base. The new Warner Robins C-5 expenditures do not seem to be registering in H036A.

Table A.4

Monthly Organic B-52 DMAG Repair Expenditures Regressed Stepwise on Monthly Fleet Flying Hours, FY97–FY00

Observations		48		
R-squared		0.26		
F		5.16		
Pr > F		0.00		
		Df	SS	
Regression		3	2.6E12	
Residual		44	7.3E12	
Total		47	9.9E12	
	Coefficient	SE	T-Statistic	P-Value
Intercept	3090856	614185	5.03	0.00
FH-3	−337.12	220.56	−1.53	0.13
FH-4	−460.44	220.45	−2.09	0.04
FH-7	−477.32	197.67	−2.41	0.02

To address this problem, our C-5 analyses run up to December 1999, not September 2000 as was the case for other systems.

Table A.9 finds no monthly lags with a significant relationship between C-5 DMAG official organic repair expenditures and C-5 fleet flying hours.

Table A.10 finds only a negative and insignificant relationship between two-month lagged C-5 fleet hours and official C-5 DMAG organic repair expenditures.

Table A.11 finds no statistically significant F-15 coefficients.

Indeed, the stepwise procedure degenerated for the F-15. No independent variables beyond the intercept term met the criterion for inclusion. Such a result is evidence of the apparent irrelevance of F-15 fleet flying hours on F-15 organic DMAG repair expenditures.

Table A.12 presents F-16 results. Unlike some other systems, four months' variables have statistically significant coefficient estimates (current month, months 5, 8, and 12), but the last three estimates are negative.

Table A.5

Monthly Organic C-130 DMAG Repair Expenditures Regressed on Monthly Fleet Flying Hours, FY97–FY00

	Observations	48		
	R-squared	0.19		
	F	0.63		
	Pr > F	0.81		
		Df	SS	
	Regression	13	9.9E13	
	Residual	34	41.2E13	
	Total	47	51.2E13	

	Coefficient	SE	T-Statistic	P-Value
Intercept	−14820582	17319230	−0.86	0.40
Current month	897.72	501.71	1.79	0.08
FH-1	133.36	472.00	0.28	0.78
FH-2	0.73	469.16	0.00	1.00
FH-3	96.43	484.42	0.20	0.84
FH-4	−292.61	507.08	−0.58	0.57
FH-5	569.77	515.10	1.11	0.28
FH-6	238.87	526.88	0.45	0.65
FH-7	−102.17	513.12	−0.20	0.84
FH-8	57.48	501.11	0.11	0.91
FH-9	13.45	502.28	0.03	0.98
FH-10	175.45	502.41	0.35	0.73
FH-11	−12.54	500.34	−0.03	0.98
FH-12	−703.97	504.90	−1.39	0.17

The stepwise finding (Table A.13) for the F-16 is analogous. More months are included than was true for other MDs, but all but the current month's point estimate are negative.

OTHER SYSTEMS' LINK BETWEEN ACCRUED DMAG REPAIR EXPENDITURES AND FLEET FLYING HOURS

We replicated the preceding analyses using accrued, rather than official, DMAG repair expenditures. We again undertook full (current month and 12 monthly lags) and stepwise regression. We also estimated a "flying hour sum" regression in which only a constant term and the sum of the current and 12 monthly lags of flying hours are independent variables. As noted in Chapter Four, Greg Hildebrandt

Table A.6

Monthly Organic C-130 DMAG Repair Expenditures Regressed Stepwise on Monthly Fleet Flying Hours, FY97–FY00

	Observations	48		
	R-squared	0.05		
	F	2.36		
	Pr > F	0.13		
		Df	SS	
	Regression	1	2.5E13	
	Residual	46	48.7E13	
	Total	47	51.2E13	
	Coefficient	SE	T-Statistic	P-Value
Intercept	−1096530	4211147	−0.26	0.80
FH-5	334.19	217.76	1.53	0.13

suggested this additional estimation to assuage concerns about highly colinear independent variables.

In the interest of parsimony, Table A.14 presents the three types of results for cargo/tanker MDs (C-130, C-135, C-141, and C-5). The FH coefficient sum is the sum of the 13 flying hour coefficient estimates. The asterisks show when the sums are statistically significantly different from zero.

We would like to highlight two points in Table A.14. First, the four systems' results are different on a number of dimensions: regression R-squared, monthly lags chosen in the stepwise regression, and the sign and magnitude of the intercept terms.

What unifies the four cargo/tanker systems is that each shows a net positive effect, under all three estimation procedures, of flying hours on accrued organic DMAG repair expenditures. Only in some of the cases, however, is the net effect estimate statistically significantly positive. In general, the findings suggest (although the timing and magnitude of the effects vary) that increasing cargo/tanker flying hours eventually results in more DMAG expenditures.

The smaller scale of the "flying hour sum" coefficient estimates is not a surprise. In the first two regressions, each regression coefficient

Table A.7

Monthly Organic C-141 DMAG Repair Expenditures Regressed on Monthly Fleet Flying Hours, FY97–FY00

Observations		48		
R-squared		0.29		
F		1.08		
Pr > F		0.41		
		Df		SS
Regression		13		9.2E12
Residual		34		22.3E12
Total		47		31.5E12
	Coefficient	SE	T-Statistic	P-Value
Intercept	−19541	653263	−0.03	0.98
Current month	−564.23	288.30	−1.96	0.06
FH-1	292.10	332.47	0.88	0.39
FH-2	84.72	315.07	0.27	0.79
FH-3	173.96	295.80	0.59	0.56
FH-4	−347.17	298.11	−1.16	0.25
FH-5	97.12	299.09	0.32	0.75
FH-6	336.54	282.53	1.19	0.24
FH-7	167.64	306.70	0.55	0.59
FH-8	−676.55	334.22	−2.02	0.05
FH-9	342.04	357.95	0.96	0.35
FH-10	30.69	344.69	0.09	0.93
FH-11	19.79	322.97	0.06	0.95
FH-12	182.55	257.19	0.71	0.48

represents the total effect of an increase of one flying hour on DMAG expenditures. The "flying hour sum" coefficient, by contrast, represents the one-month estimate of an effect that will, in total, last for 13 months (the current month through 12 months from now, inclusive). Roughly speaking, we expect the FH coefficient sums to be 13 times larger than the "flying hour sum" coefficients. To adjust for this scale difference, Table 4.5's "constrained regression" coefficient estimates are 13 times the "flying hour sum" coefficients shown here.

As shown in Table A.15, the opposite finding holds for bombers and fighters. Each of these MDs has a negative coefficient estimate of the cumulative effect of fleet flying hours on accrued organic DMAG repair expenditures. Only in some of the cases, however, are the net coefficient estimates statistically significantly different from zero.

Table A.8

Monthly Organic C-141 DMAG Repair Expenditures Regressed Stepwise on Monthly Fleet Flying Hours, FY97–FY00

		Df	SS	
Observations		48		
R-squared		0.11		
F		5.75		
Pr > F		0.02		
		Df	SS	
Regression		1	3.5E12	
Residual		46	28.0E12	
Total		47	31.5E12	

	Coefficient	SE	T-Statistic	P-Value
Intercept	320991	533690	0.60	0.55
FH-6	145.06	60.48	2.40	0.02

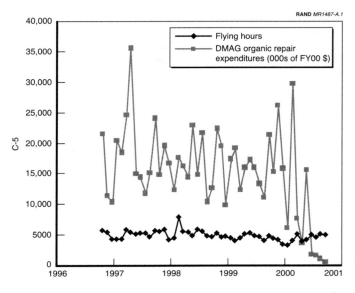

Figure A.1—C-5 Flying Hours and DMAG Organic Repair Expenditures

Table A.9

Monthly Organic C-5 DMAG Repair Expenditures Regressed on Monthly Fleet Flying Hours, FY97–December 1999

	Observations		39	
	R-squared		0.30	
	F		0.82	
	Pr > F		0.63	
			Df	SS
	Regression		13	3.1E14
	Residual		25	7.3E14
	Total		38	10.4E14
	Coefficient	SE	T-Statistic	P-Value
Intercept	15048044	17293789	0.87	0.39
Current month	1989.64	1326.31	1.50	0.15
FH-1	−830.10	1381.62	−0.60	0.55
FH-2	−2039.65	1331.21	−1.53	0.14
FH-3	−283.29	1340.47	−0.21	0.83
FH-4	−1523.72	1335.85	−1.14	0.26
FH-5	1143.22	1368.99	0.84	0.41
FH-6	1463.44	1381.39	1.06	0.30
FH-7	315.45	1358.94	0.23	0.82
FH-8	1439.26	1319.45	1.09	0.29
FH-9	−500.63	1309.75	−0.38	0.71
FH-10	−1127.90	1307.92	−0.86	0.40
FH-11	−994.59	1402.51	−0.71	0.48
FH-12	1358.81	1394.61	0.97	0.34

Table A.10

Monthly Organic C-5 DMAG Repair Expenditures Regressed Stepwise on Monthly Fleet Flying Hours, FY97–December 1999

	Observations	39		
	R-squared	0.08		
	F	3.14		
	Pr > F	0.08		
		Df	SS	
	Regression	1	0.8E14	
	Residual	37	9.6E14	
	Total	38	10.4E14	

	Coefficient	SE	T-Statistic	P-Value
Intercept	27414613	5806349	4.72	0.00
FH-2	−1996.32	1127.32	−1.77	0.08

Table A.11

Monthly Organic F-15 DMAG Repair Expenditures Regressed on Monthly Fleet Flying Hours, FY97–FY00

Observations	48
R-squared	0.17
F	0.52
Pr > F	0.89

	Df	SS
Regression	13	1.7E14
Residual	34	8.4E14
Total	47	10.1E14

	Coefficient	SE	T-Statistic	P-Value
Intercept	36889295	28021708	1.32	0.20
Current month	−276.13	694.82	−0.40	0.69
FH-1	249.50	546.98	0.46	0.65
FH-2	−456.75	560.78	−0.81	0.42
FH-3	−369.06	559.49	−0.66	0.51
FH-4	−43.06	557.15	−0.08	0.94
FH-5	330.93	556.92	0.59	0.56
FH-6	−429.94	551.22	−0.78	0.44
FH-7	−381.70	542.67	−0.70	0.49
FH-8	−119.40	542.79	−0.22	0.83
FH-9	−239.95	555.34	−0.43	0.67
FH-10	−3.30	555.98	−0.01	1.00
FH-11	−837.79	549.37	−1.53	0.14
FH-12	535.59	700.16	0.76	0.45

Table A.12

Monthly Organic F-16 DMAG Repair Expenditures Regressed on Monthly Fleet Flying Hours, FY97–FY00

	Observations	48		
	R-squared	0.58		
	F	3.64		
	Pr > F	0.00		
		Df	SS	
	Regression	13	1.6E14	
	Residual	34	1.1E14	
	Total	47	2.7E14	
	Coefficient	SE	T-Statistic	P-Value
Intercept	57265168	10987197	5.21	0.00
Current month	372.36	162.90	2.29	0.03
FH-1	−226.72	125.51	−1.81	0.08
FH-2	−251.24	131.88	−1.91	0.07
FH-3	−6.72	130.59	−0.05	0.96
FH-4	−29.28	138.71	−0.21	0.83
FH-5	−498.21	141.93	−3.51	0.00
FH-6	55.52	140.31	0.40	0.69
FH-7	87.27	138.02	0.63	0.53
FH-8	−539.78	136.80	−3.95	0.00
FH-9	−178.05	133.11	−1.34	0.19
FH-10	118.64	131.33	0.90	0.37
FH-11	−189.28	126.44	−1.50	0.14
FH-12	−506.73	163.62	−3.10	0.00

Table A.13

Monthly Organic F-16 DMAG Repair Expenditures Regressed Stepwise on Monthly Fleet Flying Hours, FY97–FY00

Observations		48		
R-squared		0.55		
F		7.00		
Pr > F		0.00		
		Df	SS	
Regression		7	1.5E14	
Residual		40	1.2E14	
Total		47	2.7E14	
	Coefficient	SE	T-Statistic	P-Value
Intercept	55928745	7985698	7.00	0.00
Current month	373.46	145.06	2.57	0.01
FH-1	−186.90	107.44	−1.74	0.09
FH-2	−296.74	105.05	−2.82	0.01
FH-5	−426.34	104.74	−4.07	0.00
FH-8	−517.77	100.82	−5.14	0.00
FH-11	−209.96	103.94	−2.02	0.05
FH-12	−481.66	153.43	−3.14	0.00

Table A.14

Cargo/Tanker Aircraft Accrued Expenditure—Flying Hour Estimations, FY97–FY00

	C-130	C-135	C-141	C-5
Full R-squared	0.3986	0.3600	0.5852	0.4043
Full F	1.68	1.43	3.58	1.25
Pr > F	0.1121	0.1988	0.0015	0.3051
Intercept	−21093207	404572	−265733	10668207
Current month	1318**	11	9	527
FH-1	205	−23	204	738
FH-2	190	−14	−157	−856
FH-3	−130	11	−121	416
FH-4	303	−18	126	1128
FH-5	169	71	40	−737
FH-6	149	−80	−67	−346
FH-7	−17	24	91	238
FH-8	279	1	−59	826
FH-9	83	7	−104	−752
FH-10	−217	45	102	1084
FH-11	−111	9	−141	−910
FH-12	−820*	32	271*	−167
FH coefficient sum	1401	76	194**	1189
Stepwise R-squared	0.3213	0.2358	0.5153	0.1713
Stepwise F	6.79	6.79	23.39	3.62
Pr > F	0.0008	0.0027	0.0001	0.0374
Intercept	−7335033	21657	−321397	6536509
Current month	1325**			
FH-1				
FH-2				
FH-3				
FH-4	408*			1050
FH-5				
FH-6				
FH-7				
FH-8				893
FH-9			−89	
FH-10		65**		
FH-11				
FH-12	−1050**	35	276**	
FH coefficient sum	683	100**	187**	1943*
Flying sum R-squared	0.0175	0.0077	0.3853	0.0324
Flying sum F	0.80	0.35	28.21	1.21
Pr > F	0.3752	0.5571	0.0000	0.2795
Intercept	−8156751	962595	−136989	7226919
Flying hour sum	53	3	14**	139

NOTE: * denotes a coefficient estimate that is statistically significant at the 95% confidence level; ** denotes 99% significance.

Table A.15

Bomber/Fighter Aircraft Accrued Expenditure—Flying Hour Estimations, FY97–FY00

	B-1	B-52	F-15	F-16
Full R-squared	0.3328	0.5410	0.1900	0.7495
Full F	1.27	2.99	0.60	7.60
Pr > F	0.2810	0.0054	0.8401	0.0001
Intercept	1894771*	6057260**	37530593	55022814**
Current month	274	−385	−336	42
FH-1	27	−311	57	−215**
FH-2	100	−335	−225	−89
FH-3	−30	−169	−571	−120
FH-4	−207	−307	−2	−215**
FH-5	−42	−394*	337	−174*
FH-6	−38	−368	−472	−143*
FH-7	−110	−341	−310	−107
FH-8	−168	−254	−300	−132
FH-9	−47	−128	−249	−238**
FH-10	−12	110	275	−97
FH-11	−97	−41	−910	−104
FH-12	−343*	44	617	−118
FH coefficient sum	−693	−2879**	−2089	−1710**
Stepwise R-squared	0.2887	0.4739		0.6579
Stepwise F	4.26	6.00		12.82
Pr > F	0.0055	0.0002		0.0001
Intercept	1767400**	5069030**		46558775**
Current month	257	−479**		
FH-1				−259**
FH-2		−333		
FH-3				
FH-4	−257*	−292		−282**
FH-5		−480**		−243**
FH-6		−346*		
FH-7		−412*		
FH-8	−277*			−208**
FH-9				−294**
FH-10				
FH-11				
FH-12	−355**			−128*
FH coefficient sum	−632*	−2342**		−1414**
Flying sum R-squared	0.1514	0.2738	0.0234	0.5216
Flying sum F	8.03	16.97	1.08	49.07
Pr > F	0.0069	0.0002	0.3045	0.0000
Intercept	2541950	4720745	32037630	50716508
Flying hour sum	−79**	−164**	−132	−120**

NOTE: * denotes a coefficient estimate that is statistically significant at the 95% confidence level; ** denotes 99% significance.

THE PERILS OF CROSS-SYSTEM ANALYSIS

Appendix A demonstrated that there is little consistent, cross-system, short-term relationship between fleet flying hours and DMAG repair expenditures. We note in this appendix, however, that if one aggregates flying hour and expenditure data across systems, one can find a spurious positive and significant relationship between fleet flying hours and DMAG repair expenditures.

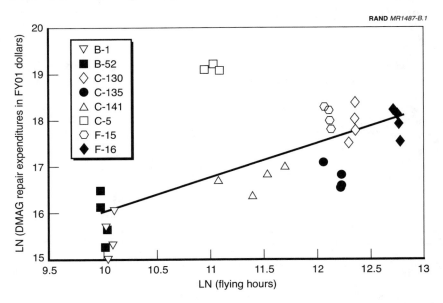

Figure B.1—Eight MDs' Flying Hours and DMAG Repair Expenditures

To illustrate this phenomenon, Figure B.1 plots the natural log of annual official DMAG organic repair expenditures against the natural log of annual fleet flying hours. There are 31 observations in Figure B.1: four fiscal years (FY97–FY00) for seven MDs (B-1, B-52, C-130, C-135, C-141, F-15, and F-16) and three fiscal years of the C-5 (we omitted FY00 because of data problems).

The regression line in Figure B.1 plots the regression findings shown in Table B.1.

How is it, then, that data that show no consistent flying hour influence on DMAG repair expenditures show a positive and significant impact of the natural log of flying hours on the natural log of DMAG expenditures when analyzed across systems?

The explanation, we believe, is that cross-system data aggregate two systems that do not fly a large number of hours and do not cost an enormous amount of money (the B-1 and B-52) with three systems that both fly more and cost more in aggregate (the C-130, F-15, and F-16). The C-5 is the high-center outlier in Figure B.1: C-5s cost a surprisingly large amount given their number of flying hours.

Table B.1

Multisystem Regression of Natural Log of Annual Organic DMAG Repair Expenditures Regressed on Natural Log of Annual Fleet Flying Hours, FY97–FY00

	Observations		31	
	R-squared		0.38	
	F		17.95	
	Pr > F		0.00	
			Df	SS
	Regression		1	15.593
	Residual		29	25.193
	Total		30	40.786
	Coefficient	SE	T-Statistic	P-Value
Intercept	8.9820	1.9442	4.62	0.00
LN(flying hours)	0.7139	0.1685	4.24	0.00

Our verdict is that cross-system estimations of this sort are spurious in suggesting a positive and significant linkage between fleet flying hours and DMAG repair expenditures.

COSTS, PRICES, AND BUDGETING FACTORS: BACKGROUND NOTES

It is easy to confuse concepts like "cost," "price," and "budgeting factor." This appendix reviews these concepts, distinguishes them, and explains how they relate to the work in the text.

SOME BASIC DEFINITIONS

A "cost" is the value of something that an organization gives up when it does something: The activity of depot maintenance imposes a cost on the Air Force by consuming real resources that the Air Force has the right to use and that it cannot use elsewhere when it decides to use these resources in depot maintenance. Two kinds of costs are important to think about. One is a cost when the Air Force pays to get or use an asset or service from someone else. For example,

- The Air Force pays its workers wages and benefits in exchange for their services.

- The Air Force pays a contractor or other government organization to repair a component.

- The Air Force buys a spare part or equipment or pays for the construction of a facility.

In each case, money passes from Air Force accounts to external parties. The second cost is the opportunity cost of using assets that the Air Force owns or controls the rights to—it has already paid for the rights to use. For example,

- The Air Force uses a test stand to repair one item rather than another.

- The Air Force asks a full-time employee to spend her time repairing an item rather than going to a training class.

In each case, the Air Force experiences no direct cash flow, but it uses up a valuable asset under its control that it could have applied in another way.

These costs are the real costs relevant to planning when the Air Force makes investments, when it uses the assets it has invested in, and when it buys additional goods or services. When it makes investments, it wants to consider the money cost of the investment. When it uses assets it has invested in, it wants to consider their value of alternative uses. When it buys additional goods and services, it wants to consider the money cost of these purchases.

A "price" is a transfer between two parties. One party pays the price and experiences it as a cost, because it could have used the money spent to buy something else. The other party receives the price and experiences it as income. This income gives the seller resources it can use to buy things it needs.

The Air Force uses internal transfer prices to transfer funds between its operating commands and AFMC. Operating commands pay prices to the SMAG for spares and component repair services. Operating commands pay prices to the DMAG for programmed depot maintenance and engine overhauls. DMAG pays prices to SMAG for parts; SMAG pays DMAG prices for component repair. All of these prices are simply transfers from one Air Force fund to another; in themselves, they impose no real costs on the Air Force as a whole and generate no income for the Air Force as a whole. Rather, each use of a price imposes a cost on one Air Force organization and generates income for another Air Force organization.

As long as the operating commands must turn to AFMC for support, DMAG must turn to SMAG, and SMAG must turn to DMAG, these internal transfer prices do not create the high-powered incentives typically attributed to prices in a free market. The existence of these prices typically motivates buyers and sellers to seek ways to escape the locked relationships they have with one another. But as long as the players are locked together, such prices function primarily to transfer funds from one administrative claimant to another.

A "budgeting factor" is a parameter the Air Force uses to construct future budgets for specific Air Force organizations. For example, as defined above, DMAG needs a budget to pay all the external parties who provide assets and other goods and services that the Air Force needs for depot maintenance. SMAG needs a budget to pay the external parties who supply assets and other goods and services that the Air Force needs to procure reparable and consumable parts. Each of these organizations develops its own budget, anticipating what payments it will have to make to provide the depot-level maintenance and supply support the Air Force needs. But the Air Force does not fund these organizations directly. Rather, the Air Force funds its operating commands, which then administratively transfer funds to DMAG and SMAG through prices. The Air Force uses budgeting factors tied to key "cost drivers" in the operating commands to construct the commands' budgets. Keep in mind that the "costs" in question are not costs to the Air Force—which the DMAG and SMAG must cover each year—but costs to the operating commands in the form of the prices they pay for DMAG and SMAG services. DMAG and SMAG then make decisions that seek to generate enough income to cover the budgets they have developed to define their expected costs during any year.

For simplicity, the discussion above says nothing about timing. Timing can be important because

- The Air Force records an event—for example, a payment—at a different time than it actually occurs, creating uncertainty about actual timing.

- Discount rates and inflation rates affect the meaning of costs, prices, and budgeting factors stated for different dates.

- An announcement at one date can affect actions at another date, transferring the time at which an organization experiences the effects of those actions. For example, if an organization obligates money today or even expects today to obligate money in the future, it may experience elements of the "real cost" associated above with an actual expenditure, which could come at a much later date. Similarly, a proposed budget may lead an organization to act before authorizations, appropriations, and a final budget define the money it will actually have available.

These timing factors can be critical. They need to be considered in the particular circumstances of any policy issue.

Costs, prices, and budgeting factors all have incentive effects. Costs by definition constrain any organization's alternatives. An organization naturally asks whether the value it gets from a good, services, or decision exceeds its cost. If value is not high enough, the organization must ask why it needs the good or service or why it should make the decision.

By transferring money, prices impose costs on the buyer and generate income for the seller. Economists often argue that prices should be designed to align the incentives faced by buyer and seller; the price should reflect the real cost to the seller, so that the buyer agrees to pay the price only when the value of the transaction to itself exceeds the true cost of the transaction to the seller. When buyer and seller are locked together, such decisions cannot occur. When prices are not meant to inform decisions, it is best to design them for administrative simplicity and to coordinate them with budgeting factors to ensure that the financial management system does not artificially constrain the substantive performance of the supply chain. Such actions in effect seek to wash maladaptive incentives out of an internal transfer pricing system.

Over the long turn, budgeting factors can have incentive effects if the organization being funded asks how a budgeting factor compares with the cost it expects to perform the activity in question. For example, an operating command could try to increase flying hours if the budgeting factor it received for flying hours were too high and to cut flying hours if the factor were too low. Any incentives associated with budgeting factors obviously operate on a different time horizon than the incentives associated with costs or prices discussed above.

OUR FOCUS

Our focus is specific. Given the real costs that AFMC can expect to incur in a particular year to support the operating commands, defined in terms of real payments to external parties for assets and other goods and services, how well does the current budgeting method to cover these costs work? And more specifically, how well does that method work with regard to the real costs in the DMAG? Phrased in this way, these questions raise the following subsidiary questions:

1. What methods does the Air Force use to budget for these costs?

2. How does the money budgeted move from the Air Force organizations that receive the money to the Air Force organizations that pay the Air Force's bills?

3. How should we measure the "real costs" to the Air Force of paying its bills?

4. How can we link answers to the three subsidiary questions above to answer our primary questions?

1. The Air Force process to build the budgets that will cover DMAG costs is very complex. At the heart of this process are two cost drivers and associated budgeting factors:

- Number and types of programmed depot maintenance, engine overhaul, and other major actions. The cost drivers are the number and type; the budgeting factors are costs to the operating commands of receiving these services from AFMC.

- Number and types of flying hours. The cost drivers are the number and type; the budgeting factors are the costs to the operating commands of receiving the services required from AFMC to support these flying hours.

For simplicity, we focus on the second type of budgeting factor by looking only at the costs that DMAG associates with JONs that service day-to-day operating command demands likely to be driven by flying hours.

2. The Air Force uses prices for individual component repairs to move funds budgeted for the operating commands to the SMAG; it uses another set of prices to move funds from the SMAG to the DMAG. The SMAG and DMAG both operate under financial management constraints that seek to make their incomes in any fiscal

year as close as possible to their costs. That is, each maintains a portion of the Air Force Working Capital Fund. And each seeks to maintain a net operating result (NOR) of zero in each fiscal year of operation. When this nominal goal is achieved, all of the money budgeted for operating commands to cover their flying hours each year flow through the SMAG to the DMAG, which then spends all of that money to pay external sources of goods and services. Expenditures need not occur in the same year as the budget. Budgeted funds this year can cover expenditures next year; budgeted funds from previous years can cover expenditures this year. But the DMAG and SMAG both operate their working capital funds within very tight tolerances; the more closely budgeted funds and expenditures match one another in real time, the more smoothly these working capital funds run.

3. The "real costs to the Air Force" relevant to this inquiry are the costs that the Air Force has budgeted for to pay for the AFMC services to support day-to-day operations in the operating commands. Ideally, we would identify all cash payments from AFMC required to support the services provided in the DMAG and SMAG. The AFMC financial management community maintains such data, but it does not link them to information directly relevant to the operating commands because it has not been asked to do so. Hence, we sought data on "real costs" as close to this ideal as possible to link to operating command demands.

A search through alternatives in AFMC revealed the H036A database as the best compromise among these goals. Most important, it provides a link between expenditures in DMAG shops and identifiable weapon systems in the operating commands. It identifies the cost that AFMC allocates to each JON each month. Although it does not take us directly to the checks that the Air Force cuts to pays its bills in DMAG, it gets close. The following details help explain how close it gets.

A JON covers a particular maintenance action in a particular resource cost recovery center. AFMC does not maintain data on the actual checks cut to support each individual JON. At best, it can maintain expenditures to external parties at the RCCC level; in some cases, it maintains such expenditure data for several RCCCs managed jointly. AFMC policy uses a set of formulas to allocate expenditures to individual JONs.[1]

- One formula uses a standard for the labor hours required to achieve a task to allocate standard labor hours to a JON. This standard is based on historical experience with each task and is updated periodically as experience accumulates.

- Another formula sums standard hours across JONs to determine how many standard labor hours have been expended at the RCCC level or higher.

- AFMC uses the ratio between this total standard figure and the total actual labor hours spent on all work at this level to define a measure called "labor efficiency"; the higher the labor efficiency, the more standard hours a given level of actual hours generates.

- AFMC divides its actual cost per labor hour by its labor efficiency to generate a value for the labor cost per standard hour.

- To identify the labor cost of a JON, AFMC can then multiply the standard hours for the JON by the calculated cost per standard hour.

This cost includes a factor to capture the cost of low-value consumables and other materiel. The cost of higher-value materiel is identified directly with each relevant JON. Indirect costs are added with a set of standard factors.

Each JON "expenditure," then, includes

- an estimated cost for labor that, summed across all labor employed in the relevant accounting unit, covers all actual labor costs in that unit,

- a factor to pick up the estimated cost of low-value materiel in proportion to standard labor applied,

- actual cost for high-value materiel, and

---

[1]An OSD study is reviewing how data enter H036A to determine whether this policy is applied in practice. Because the study was not complete when we published, we cannot look beyond the formal policy itself.

- a charge for indirect costs that covers all other costs in the DMAG.

How close is this estimate of expenditures to the actual value of checks cut in a particular period? Two questions are important:

- How close are the actual and the attributed labor cost? As noted above, the sum of attributed costs finally equals the actual cost as you aggregate more JONs. We aggregate JONs to develop a value for all JONs associated with a particular weapon system (at the MD level). This aggregation alone will tend to wash out random errors associated with individual JONs. That said, any remaining error will weaken any relationship we can establish between expenditures and flying hours, and we cannot quantify how large that effect might be. Remaining error will be larger and more troubling if systematic biases exist in the standard labor hours associated with one MD when compared with the standard labor hours associated with another MD drawing services from DMAG shops within the same accounting center. If such biases exist, the nominal method explained above systematically overestimates the actual labor hours for one MD and underestimates the actual labor hours for another. We are unaware of any such biases, and we have not been given any reason to expect them. Further work may be appropriate to determine their influence.

- How close in time are the expenditures attributed to a JON and the actual government checks associated with these expenditures? This is the subject of an ongoing OSD study; no results were available when we published. We assume that AFMC attributes expenditures to JONs close to the time when the Air Force pays for the goods and services relevant to these JONs. Differences could introduce errors that weaken any relationship we can establish between expenditures and flying hours, and we cannot quantify how large that effect might be.

4. We link Air Force budgeting methods to "real costs" to the Air Force by crafting an analytic method that reflects all of the above considerations.

- We focus on flying hours as the primary cost driver, because the Air Force uses flying hours to budget for the support of day-to-day operations.

- We employ a definition of "real costs" that allows us to use existing data to link costs to operating individual weapon systems day-to-day.

- We focus on the expenditures that that definition allows us to link most directly: DMAG expenditures on JONs that can be associated with the repair of components linked to specific weapon systems.

- We recognize difficulties in the potential gap—in the definition of labor costs and in timing—between measured expenditures on JONs and the ideal "real cost" that we seek. We expect such difficulties to affect primarily the error term in our statistical model.

- We adjust the Air Force data to allocate the cost of each JON to a month. For our purposes, subject to the concerns about data problems, we believe this allocation yields a more meaningful measure of real cost than the Air Force measure based on JONs when they close out.

- We depend on management decisions in the operating commands, SMAG, and DMAG, which are consistent with each of their organizational goals, to transform demands associated with flying hours into DMAG expenditures.

- Recognizing the uncertainties about the levels and timing of effects implied by the above considerations, we use an open definition of a lag structure in our statistical model. We have no basis for closing that definition or even expecting that it will look similar across weapon systems. We want the data to drive any relationship we can find between flying hours and DMAG expenditures as hard as possible.

Stated in terms of costs, prices, and budgeting factors, our analysis:

- Tries to focus on real *costs* to the Air Force of supporting day-to-day operations in the operating commands.

- Has no direct interest in the values of *prices* that act as simple transfers within the Air Force. But it recognizes that the current use of such prices, with associated working capital funds and controls on these funds, should play a critical role in how demands in the operating commands affect real costs to the Air Force. It takes those effects as given and focuses on understanding how well the current system links operational budgets and actual costs to the Air Force in any year.

- Takes the *budgeting factors* that the Air Force uses to reflect the cost of day-to-day operations as given. Very simply, it asks how well these factors work to generate budgets for the operating commands that will in fact cover the real costs to the Air Force that these commands impose in their day-to-day operations.

Adams, John L., John B. Abell, and Karen Isaacson, *Modeling and Forecasting the Demand for Aircraft Recoverable Spare Parts*, RAND, R-4211-AF/OSD, 1993.

Air Force Materiel Command, *Fiscal Year 2000 Annual Report.*

Air Force Materiel Command Reparable Spares Management Board (Frank Camm, chair), *Final Report*, Wright-Patterson Air Force Base, Ohio, March 1998.

Bachman, Tovey C., and Karl Kruse, *Forecasting Demand for Weapon System Items*, Logistics Management Institute, McLean, VA, DL310R1, July 1994.

Baldwin, Laura H., and Glenn A. Gotz, *Transfer Pricing for Air Force Depot-Level Reparables*, RAND, MR-808-AF, 1998.

"Boeing to Examine Kelly AFB Facilities at San Antonio for Future Work," Boeing news release 97-179 (http://www.boeing.com/news/releases/1997/news_release_970908n.html), September 8, 1997.

Brauner, Marygail K., Ellen M. Pint, John R. Bondanella, Daniel A. Relles, and Paul Steinberg, *Dollars and Sense: A Process Improvement Approach to Logistics Financial Management*, RAND, MR-1131-A, 2000.

Brown, Bernice B., "The Unpredictability of Demand for Aircraft Spare Parts," RAND, internal document, 1956.

Camm, Frank, and H. L. Shulman, *When Internal Transfer Prices and Costs Differ: How Stock Funding of Depot-Level Reparables Affects Decision Making in the Air Force*, RAND, MR-307-AF, 1993.

Cohen, I. K., John B. Abell, and T. Lippiatt, *Coupling Logistics to Operations to Meet Uncertainty and the Threat (CLOUT)*, RAND, R-3979-AF, 1991.

Crawford, Gordon, *Variability in the Demands for Aircraft Spare Parts: Its Magnitude and Implications*, RAND, R-3318-AF, 1988.

Feinberg, Amatzia, H. L. Shulman, L. W. Miller, and Robert S. Tripp, *Supporting Expeditionary Aerospace Forces: Expanded Analysis of LANTIRN Options*, RAND, MR-1225-AF, 2001.

Gillert, Douglas J., "Closing Kelly: Three Priorities," *American Forces Information Service News Articles*.

Hodges, James S., *Modeling the Demand for Spare Parts: Estimating the Variance-to-Mean Ratio and Other Issues*, RAND, N-2086-AF, 1985.

Keating, Edward G., and Susan M. Gates, *Defense Working Capital Fund Pricing Policies: Insights from the Defense Finance and Accounting Service*, RAND, MR-1066-DFAS, 1999.

Pyles, Raymond A., and Hyman L. Shulman, *United States Air Force Fighter Support in Operation Desert Storm*, RAND, MR-468-AF, 1995.

Robbert, Albert A., Susan M. Gates, and Marc N. Elliott, *Outsourcing of DoD Commercial Activities: Impacts on Civil Service Employees*, RAND, MR-866-OSD, 1997.

United States Department of Commerce, Bureau of Economic Analysis Web site (http://www.bea.doc.gov).

United States General Accounting Office, *Air Force Supply Management: Analysis of Activity Group's Financial Reports, Prices, and Cash Management*, GAO/AIMD/NSIAD-98-118, Washington, DC, June 1998.

United States General Accounting Office, *Air Force Supply: Management Actions Create Spare Parts Shortages and Operational Problems*, GAO/NSIAD/AIMD-99-77, Washington, DC, April 1999.

United States General Accounting Office, *Air Force Depot Maintenance: Analysis of Its Financial Operations*, GAO/AIMD/NSIAD-00-38, Washington, DC, December 1999.

United States General Accounting Office, *Air Force Depot Maintenance: Budgeting Difficulties and Operational Inefficiencies*, GAO/AIMD/NSIAD-00-185, Washington, DC, August 2000.

Wallace, John M., Dale A. Kem, and Caroline Nelson, *Another Look at Transfer Prices for Depot-Level Reparables*, Logistics Management Institute, PA602T1, McLean, VA, January 1999.